POTENT TRACE MINERAL

Once considered just another trace mineral ingested in food and drink, silica has recently been shown to play a powerful role in promoting overall good health and preventing early aging. Cutting-edge research shows that silica helps heal wounds and build bone and connective tissue, particularly collagen. It also helps prevent cardiovascular and Alzheimer's diseases. When taken orally as a supplement, it can turn back the clock as it revitalizes skin, hair and nails.

Edward A. Lemmo, Ph.D. is a nutritionist in private practice and president of Lemmo & Associates, a nutrition and health consulting group in New York City. He earned a doctorate in nutrition from Rutgers University and is an adjunct faculty member of the Human Nutrition Program at the University of New Haven, Connecticut. Dr. Lemmo conducts seminars in nutrition and has appeared as a guest on numerous radio and television programs discussing nutrition-related topics across the country.

Silica

The mineral building block
that promotes healthy bone and
connective tissue and helps prevent
premature aging and cardiovascular
and Alzheimer's diseases

Edward A. Lemmo, Ph.D.

Keats Publishing, Inc. New Canaan, Connecticut

Silica is not intended as medical advice. Its intent is solely informational and educational. Please consult a health professional if you have questions about your health.

SILICA

ISBN: 0-87983-905-8

Printed in the United States of America

5 6 7 8 9 RCP/RCP 0 9 8 7 6

Contents

INTRODUCTION

Dr. Linus Pauling once said that "one could trace every sickness, every disease and every ailment to a mineral deficiency!" Minerals are key elements of life that play essential roles throughout the body in metabolism and maintenance of physiological systems. Minerals as nutritional elements have been divided into two categories based on the amount needed by the body. Those that are required in large amounts are classified as macrominerals. Those that are required in relatively small amounts are classified as microminerals, or trace minerals. Discovery of the requirement and role for minerals has advanced in recent years and continues to expand.

Silicon is a mineral that is not commonly written about as an essential nutrient. Yet the 1993 Nobel prize winner professor Adolf Butenandt proved decades ago that life is impossible without silicon. Indeed, we now know that silicon has played a decisive role from the beginning of evolution. In fact silicon is the second most abundant element in the Earth's crust. Subaerial weathering processes produce orthosilicic acid, which is eventually deposited in the oceans. In seawater, various groups of organisms—diatoms, radiolarians, silicoflagellates, sponges and some fungi—utilize silicon primarily as a structural component.

Silicon has also played a role in the development of world civilization. From the caves of Cro-Magnon and Altamira, where flint was used to make early tools and weapons, to today's Silicon Valley and outer space, the roles for silicon are vast. In the modern world, silicon's numerous applications will be revealed as new challenges and opportunities open up in the future. Scientists will find old and new forms of colloidal silica that will contribute to the quality of life of people around the world.

Though silicon is found in rock crystals such as quartz or flint, and is widely available in food, silicon does not occur freely in nature. Silica (silicon dioxide), the "active" form of silicon, is the natural compound formed from the elemental mineral silicon and oxygen. Since silica is the natural form of this mineral, it is the preferred term used in this book.

It is interesting to note that in Germany, where a great deal of research has been conducted on this mineral, the preferred term is silicic acid. Indeed, silicic acid is found in lakes, rivers and the world's oceans. Furthermore in many biological studies, data are given in terms of "silicon" rather than "silica." However, since there is little evidence that silicon occurs in any form other than with oxygen, the preferred term used in North America is silica.

Human beings, animals and plants have an essential need for silicon in the form of silica. It represents about 0.05 percent of our body weight. Silica is especially needed in the lungs, spleen, lymph nodes, blood and blood vessels, connective tissue, nails, hair, skin, bone, cartilage and tendons.

It wasn't until 1972 through research conducted at UCLA by Dr. Edith Carlisle that silica was recognized as an essential trace element. From that point on research has demonstrated beyond a shadow of a doubt the necessity and essential requirement of silica in fulfilling structural and metabolic needs of all plants, animals and humans.

Since Dr. Carlisle's 1972 demonstration of the essentiality of silica in higher animals, bioinorganic chemists have speculated about the site and mechanism of its action. The National Research Council currently considers silica a micromineral under investigation, as it has been since the early 1970s.

Interestingly, there is no Recommended Dietary Allowance for silica. Considering all the tissues in the human body that contain silica, one would think there would be greater interest in it and a recommendation for daily intake. Problems associated with silica deficiency are under investigation and further suggest the essentiality of silica for humans.

Because it occurs naturally in food and drink, silica is considered highly safe and virtually nontoxic. Mineral water and spa water are good sources of silica. Unfortunately, silica is easily lost in food processing.

The story of silica—its importance in nature and potential as an essential trace mineral for good health—is of great interest in the field of nutrition. Based on a review of the scientific literature, it appears silica has been an overlooked mineral. But not anymore. Silica may indeed play key roles in the body that have far-reaching consequences for improved health and well-being.

HISTORICAL USES

Silica and its natural source of raw material quartz have played an ever-increasing role in the evolution of humanity. Flint gave humans the full use of and control over fire, an advancement as significant as the development of the wheel. From earliest times, quartz crystals were used for medicinal purposes as objects of great healing power. As gemstones, quartz has adorned bodies and sacred objects since before recorded history.

As people evolved and began erecting permanent buildings, they discovered many uses for silica in such materials as glass, bricks and concrete. Silica constitutes a majority of the stone used in old and new buildings alike. It is an integral part of stucco and plaster as well. Industrially, silica has been used as an additive in rubber and plastic products, ceramics, desiccants, lubricants, antislip coatings, photographic emulsions, investment castings and soil retardants in shampoos and adhesives, to mention just a few applications. In the optical field, silica plays a role in the creation and polishing of lenses and any chips that may be involved in optical applications. Silica and silicon formed the basis for the greatest technological shift since the industrial revolution—the development of the semiconductor. This shift in technology has completely altered life as we know it, and its momentous impact has yet to be fully acknowledged and understood. Using silica as a therapeutic mineral supplement is just the most recent in a long line of useful applications.

DECADES OF RESEARCH

Louis Pasteur predicted in 1878 that someday silica would be found to be an important therapeutic substance for many diseases and would play a significant role in human health and consequently nutrition. At the beginning of the 20th century, reports from Germany and France suggested that Pasteur's prediction would become fact. These reports described therapeutic successes in treating numerous diseases, including atherosclerosis, dermatitis, hypertension and tuberculosis with sodium silicate, simple organic silica compounds and teas or extracts of silica-rich botanicals. Scientific interest in a biological role for silica dates from the early 1900s when it was believed that silica was somehow involved in the synthesis and structure of connective tissue.

However, the most substantive reports about silica's therapeutic role were announced in 1972 when Dr. Edith Carlisle reported that silica was essential for bone formation. About the same time, other reports suggested that inadequate dietary silica may contribute to some cases of atherosclerosis and hypertension, in addition to some bone disorders and the aging process. Since then, reports have periodically appeared that further support the nutritional importance of silica (Mancinella, 1991; Seaborn & Nielsen, 1993; Birchall, 1995).

Surprisingly, these reports seem to have been generally ignored or considered inconsequential by clinical and nutritional professionals, the media and the general public. Since the early 1970s, the battle to bring attention to the nutritional importance of silica has been essentially fought by Dr. Carlisle. She was the first to demonstrate that the dietary restriction of silica could induce a deficiency state in chickens. Silica-deficient chicks had stunted growth, and their internal organs were not properly developed. Dr. Carlisle concluded from these findings that silica is involved with calcium in early stages of bone calcification. Carlisle also showed that silica increases the rate of bone mineralization. Other researchers such as Schwarz and Milne in 1972 showed silica deficiency in rats resulted in slow bone growth and deformities.

Further studies by Carlisle (1974) showed that silica deficiency in the chick is associated with abnormalities involved in the formation and structure of cartilage, bone matrix and connective tissue. Since silica is a component part of ground substance (a compound of carbohydrate and protein), it is speculated that it may also play an important role in the formation of collagen. Carlisle also found silica in bone-forming cells known as osteoblasts.

Aside from its presence in osteoblasts and collagen, silica is found in blood, skin, muscle, heart, liver and thymus (Carlisle, 1974; Loeper et al., 1978). An interesting observation by these researchers showed that the silica content of these tissues decreased with age and with diseases of these tissues. Schwarz (1977) argued that a low-silica-intake level was correlated with a high incidence of atherosclerosis. His research shows silica is less present in a diseased heart, especially where calcification has occurred.

SILICA IN BIOLOGICAL SYSTEMS

Research beginning early in this century and continuing through the present time has demonstrated conclusively that silica plays a significant role in the metabolism and growth of most, if not all, living things. From the smallest diatom to the largest animal, the significance of silica is continually being demonstrated.

DIATOMS AND ALGAE

Diatoms are one of the earliest life forms to exist on this planet. Enormous deposits of diatomaceous earth (made up of skeletons of ancient diatoms) exist today, which is composed almost entirely of silica. Obviously, silica has played a long and significant role in the early development of this planet.

Siever (1957) has pointed out that the major mechanism for the precipitation of silica on the surface of the earth is biochemical. A certain minimum concentration of silica in solution is essential to the growth of each kind of diatom. Increasing silica content from 3.5 to 8.3 ppm doubles the rate of growth of one type of diatoms of which the dry weight of the cells is 4 to 22 percent silica (Lewin, 1957, 1958). However, some species which contain only 0.4 percent silica can obtain enough silica for growth from ordinary glassware.

There is as yet no evidence that silica is necessary to the metabolism of blue-green algae, but at least one type observed by Iler (1979) grows readily in concentrated 30 percent solutions of colloidal silica. Well-washed cell membranes are found to have absorbed colloidal silica. Bolyshev (1952) believed that blue-green algae decomposed soil minerals and brought silica and alumina into solution and that silica was thus made available for utilization by certain diatoms which accompanied the algae.

The critical role of silica in the early stages of development of diatoms and algae is further suggested by abnormalities in the development of *Cyclotella cryptica* (Werner, 1968). Apparently silica plays a very fundamental role in the metabolism of algae. In the absence of silica, the entire cell becomes disorganized and cannot keep on dividing according to Reimann (1965). It is possible that silica plays a role in the DNA of algae as it may do in higher organisms.

PLANTS

Rice hulls are very high in silica, and both the straw and grain of wheat contain silica. A consequence of the silica content of grain is that beer is essentially a saturated solution of silica (Stone & Gray, 1948). Assays of 14 types of beer showed 60 to 100 ppm silica, derived almost entirely from the malt husk.

The hardness and stiffness of bamboo can partly be ascribed to silica in its fibrous structure. However, such excesses of silica are produced that masses of silica gel are often found in the hollow stems. This gelatinous material containing some organic matter, known as *tabasheer* (also *tabashir* and *tabaschir*), used to be employed throughout Asia as a medicine.

Many grasses, reeds and straws owe their weather resistance to heavy impregnation with silica (Frison, 1948). That's why they're often strong enough to be used to make thatched roofs. The shiny epidermis of rattan, used for furniture, is also richly impregnated with silica. The leaves of the palmyra palm of India, used for centuries as writing paper, contain beautiful siliceous concretions.

The Equisetum genus (horsetail) contains so much silica

it was used in the kitchen as "scouring rush." According to Frison (1948) these plants were used for centuries as abrasives; one type was employed for polishing wood, another for cleaning household utensils. Pioneers also used horsetail to clean their teeth.

It appears that although silica may not be necessary to the growth of most plants, it nevertheless often seems to have important secondary effects in promoting the plant's health. For instance, a convincing body of evidence shows that the presence of silica in certain plants seems to enhance resistance to fungus disease, making the plants appear healthier. Lundie (1913) concluded that silica is not essential as a plant food but believed that it is deposited in the epidermis and provides protection against fungus diseases such as rust.

The importance of silicic acid in increasing the resistance of plants to powdery mildew fungi was demonstrated by Wagner (1940). Silicic acid favors an accumulation and better utilization of calcium, phosphorus, potassium and magnesium in the plant. It was demonstrated that in the absence of magnesium the plant did not take up silica. Only when magnesium was present did the plant absorb silica, which then could be found in the leaves and provided resistance to the fungus (*Piricularia oryzae*).

Sprecher (1911) believed that silica has an important biological function in stimulating plants to greater growth and probably plays a role in maintaining a "physiological equilibrium" in the nutritive solutions in the soils. Also, in some soils, the addition of soluble silicates increases plant growth indirectly by liberating phosphate ions adsorbed on the soil, thus increasing the total amount of available phosphate. Sreenivasan (1934) concluded that silica in the soil facilitates the uptake of phosphorus. It therefore seems clear that the addition of silica may have a nutritional effect because it displaces phosphate ions from the adsorbed condition of the soil, thus making phosphate more available for absorption by the plant. Consequently it can be concluded that higher percentages of silica in the soil produce healthier plants.

HUMANS

Organic silica is the dietary form of the mineral element silicon used by humans. Pennington (1991) provides information on the silica content of foods and diets and shows that silica levels tend to be higher in foods derived from plants than in foods from animal sources. This is because plants obtain their silica directly from the soil in which they grow. Animals are not good silica sources for humans because the silica in their bodies is not as readily available for absorption by humans as it is from plant sources. Therefore, humans cannot depend on animal foods as reliable sources of silica. Studies reported by Birchall (1995) indicate that plants grown in silica-deficient areas have lower silica content.

Foods highest in silica include cereal grains, especially oats, barley and some rice fractions (Table 1). Average daily intakes of silica probably range from about 20 to 50 mg/day, with the lower values for animal-based diets and the higher values for plant-based diets (Seaborn & Nielsen,

Table 1. Natural Sources of Silicon*

Natural Source	Silicon (ppm dry weight)
Sugar beet pulp	23,100
Rice hulls	22,500
Oat hulls	16,900
Alfalfa	12,700
Sugar cane pulp	11,300
Soybean meal	1,680
Guar gum	1,420
Pectin preparations	1,040–1,130
Wheat bran	229–1,720
Nutri soy flour	93
Soya fluff	80
Wheat flour	21

*Adapted from Schwarz (1977), *Lancet* i: 454.

1993). Pennington estimates that 60 percent of the dietary silica is from cereals and 20 percent from water, with the latter providing silicon in its most bioavailable form as silicic acid. The silica content of adult U.S. diets, based on the Food and Drug Administration's Total Diet Study model, is 19 mg/day for women and 40 mg/day for men (Seaborn & Nielsen, 1993). This data is best explained on the basis of the total caloric intake difference, assuming that men consume more food and calories than women.

We are learning more about human metabolism of silica all the time. Silica enters the body as silicic acid from such foods as cereal grains, fruits and vegetables and leaves it in that form or as polymerized hydrated silica. Apparently, the form of dietary silica determines how well it is absorbed. Seaborn and Nielsen (1993) found that silica absorption is affected by age, sex and the activity of various endocrine glands. However, the mechanisms involved in the intestinal absorption of silica are still unknown.

Body storage of silica is limited. Based on animal research, microscopic particles of solid silica from plants were found to be absorbed from the digestive tract (Baker et al., 1961). The permeability of the gut wall to particles the size of diatoms has also been demonstrated in humans. Volkenheimer (1964) showed that diatomaceous earth particles were absorbed through the intact intestinal mucosa and passed through the lymphatic and circulatory systems where they reach other tissues of the body.

Normal adult human tissues range from 50 to 1000 ppm silicon. Schroeder (1973) estimated a 132-pound (60 kg) person has about 3 g of silicon in the body. The blood silicon averages close to 5 mcg/ml and occurs mostly, or entirely, in solution as monosilicic acid. The highest silicon concentrations normally occur in the skin and its appendages. Connective tissues, including aorta, trachea, tendon, bone, skin, hair and nails, contain much of the silica that is retained in the body. The high silica content of connective tissues apparently results from the presence as an integral component of the glycosaminoglycans (a carbohydrate containing amino sugar) and the protein complexes that contribute to its structural framework. Table 2 reports the silicon content of human tissues. In human hair, skin and nails, silicon is in a hydrogel state (gelatinous hydrate of silicic acid).

Table 2. Silicon Content of Human Tissues*

Tissue	Silicon (ppm dry basis)
Muscle	18 ± 1.3
Tendon	28 ± 1.8
Aorta	41 ± 3.3
Kidney	42 ± 3.9
Nails	56 ± 2.2
Hair	90 ± 2.0
Epidermis	106 ± 2.7

*Fregert, S. (1958), *J. Invest. Dermatol.* 31: 95.

Increased urinary output of silica with increasing intake, up to fairly well-defined limits, has been demonstrated in humans. The upper limits of urinary excretion after oral administration do not seem to be set by the ability of the kidney to excrete more silica. The limits are determined by the rate and extent of silica absorption from the gastrointestinal tract into the blood. Once silica has entered the bloodstream, it passes rapidly into the urine, because the concentration in the blood remains practically constant (Jones & Handreck, 1965). Further evidence that some forms of silica, including that in food, are well absorbed is that in rats and humans urinary excretion can be a high percentage (close to 50 percent) of daily silica intake (Seaborn & Nielsen, 1993).

A continuing study of silica in the animal body has led to the conclusion that aging is accompanied by a decrease in silica. However, it is often observed that certain trace elements are found in the tissues of adults which are not found in newborns and young children and we generally accumulate minerals as we grow older. That leads to a major question: Why do human have less silica in the body as we age? According to Charnot and Peres (1971), silica metabolism is controlled by steroid and thyroid hormones. It is thought that the changes in silicon content with age could be due to the decline in hormonal activity. Or could they be due to changes in dietary habits? Diet may be a factor since many elderly people eat fewer calories than younger adults. However, children also eat fewer calories than young adults. Could it be competition from other minerals which are absorbed preferentially over silica? Mineral competition may also be a factor, since most people do not supplement their diets with silica.

At this point scientists can only provide hypothetical answers to these questions. In their continuing research, they should consider the significance of body tissue content difference when evaluating the physiological role of silica in aging and disease prevention.

MINERALS ARE THE KEYS TO LIFE

Major minerals, also known as macrominerals, are required in amounts greater than 100 mg daily, while others, known as trace minerals, have much lower requirements (Table 3). According to government guidelines and recommendations, twelve or thirteen minerals are necessary for average health, and another eight or ten are possibly needed.

Nature has developed a beautiful ecological system for mineral consumption. Soil provides minerals for plant development. Plants serve as food for animals which assimilate them, absorbing their many nutrients. Then, humans eat

Table 3. Minerals of Nutritional Significance

Major Minerals	Trace Elements
Calcium	Boron
Chloride	Chromium
Magnesium	Cobalt
Phosphorus	Copper
Potassium	Fluoride
Sodium	Iodine
Sulfur	Iron
	Manganese
	Molybdenum
	Selenium
	Silicon
	Vanadium
	Zinc

plant and animal products. Unfortunately, erosion and un-wise farming methods have made mineral-rich soil a rare commodity. The result is not only mineral-deficient plants and livestock but mineral-deficient people. This deficiency is further compounded by the way food is processed, which also leads to mineral loss.

In the human body, minerals interact with vitamins, en-zymes and many other substances in our metabolism. At least 84 minerals have been isolated from human tissues, and at least 60 are known to function in metabolic reactions. Today researchers are awakening to the importance of min-erals in total health. Minerals are actually the foundation of our biochemical life!

TYPES OF MINERALS

Minerals can be absorbed by the body in three different forms: metallic, chelated and colloidal. Metallic minerals are most commonly used in nutritional supplements, especially for the major minerals, because larger amounts are needed. However, the percentage of absorption of these minerals is generally low. Metallic minerals are considered hydrophobic (not water soluble); minerals that have been assimilated by plants are hydrophilic, making them very water soluble. Chelated minerals are a step above metallic minerals because they are more easily absorbed. The metallic mineral is attached (chelated) to an amino acid, the building block of protein, making it easier to be assimilated by the body.

COLLOIDAL MINERALS

Colloidal chemistry is a very neglected branch of science, and much research has yet to be undertaken to determine the exact function of all the various elements. Technically, a colloid is a particle or tiny substance that retains its identity and remains in suspension in a liquid medium. (The term "colloid" is derived from the Greek word for "glue.") Unlike ordinary suspensions, colloids do not settle to a noticeable degree. Because of their extremely high ratio of surface area to volume, the particles remain in uniform distribution. Different types of colloids are found in nature. Some familiar examples are emulsions (mayonnaise, milk), aerosols (fog, smoke) and sols (ink, glue). Due to their very small particle size, colloids are easily absorbed by the cells of the body. Thus *colloidal minerals* can pass directly into the cells of our bodies, providing the highest levels of absorption by the human body.

Colloids are measured by their particle size. As the particle size gets smaller, the colloid becomes more transparent. Dr. Carey Reams, a well-known biophysicist and biochemist, discovered that colloids can become so small that they can go through glass! And a variety of different minerals can exist within a single colloid.

It is possible to test for particles down to these very small sizes using the unit of measure called the *angstrom* (Å). An angstrom equal one ten-billionth of a meter (10^{-8} cm). Generally, the size of particles in a colloid range from 10 to 10,000 Å. However, with the standard method of testing, known as spectroscopy, the results are only approximate and vary from lab to lab.

COLLOIDAL SILICA

Colloidal silica or *silica sol* (terms which are usually used interchangeably) may be defined as a dispersion of silica in a liquid medium in which the particle size of the silica is within the colloidal range. This form of silica is most easily assimilated by the human body. Therefore most supplements on the market today are colloidal silica.

INTERACTION OF SILICA WITH OTHER MINERALS

Carlisle (1984) studied the interaction of silica with the trace elements zinc (Zn), copper (Cu) and iron (Fe) in serum and tissues of rats. The concentrations of silica, iron and zinc were measured in samples of blood and tissues of rats orally receiving a soluble, inorganic silica compound dissolved in their drinking water. An increase of copper concentrations in liver and aortic walls in the experimental group was observed, with the simultaneous reduction of zinc amounts in serum and all the tissue samples in the course of the experiment. The iron concentrations in the analyzed samples did not show any significant changes between the groups. The silica levels in serum and in all the examined tissues were significantly higher in the tested group. The results therefore provide evidence of silica interaction with copper and zinc, which could result in a number of metabolic process modifications, including cardiovascular benefits.

Carlisle (1984) also showed that silicic acid influences copper utilization in the rat. When rats were fed a diet which contained silica, the level of copper in blood and other tissues increased. This could be due to a favorable effect of silicic acid on gastrointestinal absorption and transport of copper or its availability in collagen and elastin proteins of connective tissue. It is important to know this since copper deficiency has been associated with such diseases as osteoporosis and other bone deformities.

Silica also plays an important role in assisting calcium in maintaining the hardness of bone and in increasing collagen for cartilage flexibility. Carlisle's studies (1984) show that fractured bones heal faster when silica is used as a supplement.

Considering the role of silicic acid and its interactions throughout the body, the regular dietary intake of silicic acid is warranted in order to maintain a favorable balance of silicic acid relative to other minerals in the body.

MULTIPLE ROLES IN THE HUMAN BODY

Slowly but surely research is showing that silica plays an enormous number of vital roles in the human body. This versatile mineral is a veritable building block for a whole range of essential functions such as building strong bones and connective tissue and healing wounds. While silica has been shown to restore health to aging skin, hair and nails, perhaps its most life-protecting functions are in promoting optimal cardiovascular health and fighting cancer.

Though the requirements for silica intake have not yet been established, silica deficiency has been observed. Most of the signs of silica deficiency from animal studies demonstrate changes in such areas as bone and connective tissue. There is much speculation that a deficiency in silica may lead to dull, lifeless hair, brittle nails, loose, sagging skin and premature wrinkles. Many researchers have postulated that silica deficiency causes several human disorders, including atherosclerosis, hypertension, osteoarthritis and the aging process. These speculations clearly indicate the need for further study of the importance of silica nutrition, especially in aging humans.

BONE BUILDING

Most people today are familiar with the term "osteoporosis," a condition in which bone mineral content and density are decreased, with a loss of organic bone elements, collagen, osteoblasts, osteoclasts and connective tissue. Though this bone-loss process is most commonly seen in older women, more than 15 million Americans, men as well as women, have a form of this bone-thinning disease. The reasons for bone loss have traditionally been lack of vitamin D, lack of available dietary calcium, a high rate of calcium excretion, a high proportion of phosphorus to calcium in the diet or other factors that enhance secretion of the parathyroid hormone.

After middle age, three factors are thought to be the most important determinants of the rate of bone loss: estrogen, physical activity and diet. It is very common to hear about a postmenopausal woman who fell and broke her hip. Did the fall cause the hip to break or did the hip break and cause the woman to fall? The finding that postmenopausal women are most susceptible to osteoporosis implies that the process of bone loss tends to quicken with a decline in estrogen secretion (Draper & Bell, 1980).

Physical activity throughout life is important for maintaining healthy bones. Diet also plays an important role, particularly with respect to the calcium/phosphorus ratio. Although we should consume equal amounts of calcium and phosphorus, the typical diet, consisting of highly processed foods and excess protein, red meats, sugar, caffeine, tobacco, alcohol and carbonated beverages, makes it impossible to have the proper ratio of calcium and phosphorus. In order to prevent bone loss, steps must be taken early in life and then we must follow them the rest of our lives.

But that is not the end of the story. New discoveries can now help us prevent bone loss. Evidence that silica is associated with calcium in an early stage of bone building was demonstrated by Carlisle (1970). Carlisle reported silica plays a vital role in the formation of the matrix of bone and connective tissue, the very location where bone originates. It was found to be uniquely localized in osteoblasts, the active building sites in new or young bone. Then as calcium is deposited on the matrix of bone, silica is replaced by calcium as the structural mineral. This research suggests that silica serves as a primary structural mineral in the formation of bone. In fact bones are rich resources of calcium, magnesium, phosphorus and silica. Average compact bone contains by weight approximately 30 percent matrix and 70 percent mineral salts. However, newly formed bone may have a considerably higher percentage of matrix in relation to mineral salts.

It is important to stress the unique connection between calcium and silica in bone formation. The initial process involves the formation of a collagen matrix. Silica is needed for this process. Once the matrix is formed, the mineral matrix, including calcium, is laid down on the collagen framework. If one considers the relationship of calcium and silica, it is not possible to form bone without either calcium or silica. They both play equally unique and essential roles in the bone development process.

WOUND HEALING

Wound healing is a complex sequence of events, beginning with tissue injury, mediated by inflammation and ending long after reepithelialization is complete. Research and controlled clinical experience have provided a thorough understanding of the entire process so that clinicians can influence the stages of healing to decrease pain, control bleeding and infection, ensure cosmetic results and speed the time for complete healing.

Age-related differences in wound healing have been clearly documented. Although wounds in the elderly can heal, they do so more slowly, and all phases are affected. The inflammatory response is decreased or delayed, as is the proliferative response. Remodeling occurs, but to a lesser degree, and the collagen formed is qualitatively different. Diseases that affect wound healing are more prevalent in the elderly and have a greater adverse effect than in young adults. Recent trials of novel therapies to enhance wound healing suggest, however, that much can be done to improve the prognosis of elderly patients.

Highly diluted solutions of silica are widely used in homeopathic medicine to treat chronic wounds, ulcers and abscesses. The therapeutic effects of homeopathic dilutions of silica on induced chronic wounds in mice were tested by Oberbaum (1992). In each experiment three or four groups of 10 mice each were treated by adding homeopathic dilutions of silica and of saline to the drinking water of the mice for 4 to 20 days. The size of the wounds was measured every second day by an objective image analysis system. The results showed that the wounds of the silica-treated animals were significantly smaller and healed faster than those in mice treated with saline. Also the therapeutic effect increased progressively with an increase in dilution of the silica (dilution of a solution is customary in homeopathy). These results show that homeopathic dilutions of silica clearly have a therapeutic effect on wound healing.

BUILDING CONNECTIVE TISSUE AND COLLAGEN

Connective tissue is the most abundant and the most widely distributed tissue in all parts of the body. Connective tissue is comprised of collagen and elastin proteins in a matrix of "ground substance," mucopolysaccharides and other minute substances. The proportion and types of these components determine the properties of the tissue—the flexibility and toughness of cartilage, the elasticity of arteries and the nature of the organic matrix that is mineralized in the formation of bone. Connective tissue performs a variety of functions, ranging from broad areas of support, protection and binding together of other tissues to the production of certain components of the blood, storage of fat and protection of the body against bacterial invasion and disease. Some of these functions depend primarily upon the activities of special connective tissue cells, while others depend more upon the intercellular substance produced by the cells, which consists of fibrous and amorphous (meaning without form) types. Their ratio and nature differ among the varieties of connective tissues in the body.

Fibrous intercellular substance includes three types of fibers: collagenic, elastic and reticular. All three are proteins in nature. Collagen, the most abundant protein in the body, is an extremely tough protein, with a rich silica content. Therefore collagenic fibers are strong and tough yet flexible and resistant to a pulling force. These fibers form the fabric of skin, ligaments and tendons. Collagen is also a major constituent of bone and cartilage.

Connective Tissue Diseases During the past decade, the study of connective tissue disease has received great attention. More than 32 million people in the United States are affected by more than 100 types of connective tissue diseases (connective tissue disease is the newer more accurate term for collagen disease). These disorders not only affect the collagen portion of connective tissue, but involve its other protein components (elastin and mucopolysaccharides) as well. Carlisle (1970) found silica to be present in all of these tissues, and Kaufmann (1992) found connective tissue diseases are often connected to silica deficiency. Consequently, the concept that connective tissue contains silica led to the idea of silica compounds as therapy in connective tissue disorders. In fact silica has a fundamental effect on connective tissue weakness and its repercussions. As human tissue consists essentially of a system of colloids, colloidal solutions of silica, such as silica gel or liquid supplements, are particularly suitable for the treatment of connective tissue weakness (Kaufmann, 1992).

Rheumatoid Arthritis: The term "arthritis" is often used interchangeably with connective tissue disease or rheumatic disease; however, this usage is not accurate. Arthritis (inflammation of a joint) is a symptom of connective tissue disease. Rheumatoid arthritis and degenerative joint disease (osteoarthritis) are the two main types of connective tissue disease, in which arthritis is the major manifestation. Due to the chronic nature of these diseases, billions of dollars in work productivity are lost each year; an additional $1 billion is spent on disability benefits. Rheumatoid arthritis affects people at any age, but it most often begins in women between 25 and 35. It affects women three times more often than men. It is believed that the disease worsens when there is physical or emotional stress.

Degenerative Joint Disease: Degenerative joint disease (DJD) is the most common type of connective tissue disease in the United States, affecting nearly 16 million people. It has been estimated that 80 percent of people over 55 years of age have some degree of degeneration. Unlike rheumatoid arthritis, DJD is not a systemic, inflammatory disease process. For this reason, the term "osteoarthritis," frequently used interchangeably with DJD, is not an accurate synonym.

Degenerative joint disease is a wear-and-tear process in which there is degeneration of cartilage with resultant formation of irregular bony overgrowths. Most frequently seen in weight-bearing joints, it is thought to be the result of prolonged mechanic stress. DJD affects women twice as often as men and seems to have a familial tendency. Although seen most often in the elderly, the disease may be associated with obesity, previous trauma, strenuous physical labor or athletics in any age group. The most common symptom of DJD is pain, which tends to worsen with activity and improve after rest. Morning stiffness may occur but last less than 30 minutes. Physical mobility may be impaired and muscle spasm is often present.

Kaufmann (1992) states that in connective tissue weakness, muscles, joints and tendons are poorly developed. For that reason sprains, dislocations, pulling, overstretching and hernias often develop. Kaufmann contends that silica has a fundamental effect on connective tissue weakness and its repercussions. Carlisle (1984) extensively documented the positive role of silica in connective tissue formation.

PROMOTING HEALTHY SKIN, HAIR AND NAILS

Fresh, natural beauty comes from within. A special kind of beauty radiates from healthy people. You see their well-being reflected in their shiny hair, clear skin and strong, healthy nails. It's the kind of beauty that cannot be obtained through the artificial use of cosmetic creams, lotions or hair conditioners.

The Structure of Skin Skin covers the entire surface of the body and therefore is the largest organ of the body. Skin performs many essential functions such as protection from sun, infection and environmental toxins, helping with temperature regulation and serving as a sense organ. Skin also produces vitamin D and plays a minor role in the elimination of water and salts. It consists of two main layers, the epidermis and the dermis. These layers are further subdivided into other functional layers.

The epidermis is the thin surface layer of the skin, which varies in thickness in different parts of the body. For example, it is thinnest on the eyelids and thickest on the palms of the hands and soles of the feet. Overexposure of the epidermis to ultraviolet light can lead to skin cancer. A more usual result of overexposure to sunlight is premature aging of the skin. This causes, among other things, wrinkles which appear on the outer surface. Part of the cause of wrinkling is due to structural changes in the underlying connective tissue fibers that support the epidermis.

The dermis is the deeper layer of the skin. It is composed of two layers of connective tissue. The dermis layer contains blood vessels, sweat glands, sense receptor cells, hair follicles and oil-producing sebaceous glands.

Beneath the dermis lies the subcutaneous tissue. Though not considered part of the skin, this tissue plays a vital role in maintaining a healthy body. Subcutaneous tissue varies in different parts of the body and in different individuals. Some is loose adipose tissue, a dense type of connective tissue. The dermis is anchored to the subcutaneous tissue by collagenic fibers. The subcutaneous tissue then attaches the dermis to underlying structures such as muscles and bones. All these layers are held together by collagen. As noted above, silica is essential for collagen production and therefore a decline in silica content in the skin leads to a weakening of the collagen structure, resulting in all the manifestations of old skin.

The Structure of Hair Hair grows from small sacs called "follicles" located just below the surface of the skin. Every follicle has its own blood supply. The thickness of hair is determined by the size of the opening in the follicle. Tiny muscles called "arrector pili" muscles are attached to each follicle and respond to stimuli. For example, when your are cold, you get goosebumps. When you are frightened, your hair stands on end. The root of the hair is surrounded by a bulb that feeds keratin, a tough fibrous protein, to each strand of hair. Sebaceous glands are located above the bulb and below the skin. These glands produce an oily substance called "sebum" which lubricates and protects the hair.

The shaft of the hair is made of three layers, the center core or medulla, the thicker middle layer called the "cortex" and the tough outer layer called the "cuticle." There are three major types of hair: straight, wavy and kinky or woolly. When seen in cross section under a microscope, straight hair is round, wavy hair is oval, and kinky or woolly hair is elliptical or kidney shaped. Hair color depends upon the quality and quantity of pigment (melanin) in the cortex (outer layer). There is a constant natural loss of hair throughout life.

The Structure of Nails Nails are primarily composed of keratin as well. The same tough protein as in hair is produced by the cells under the nail. Fingernails and toenails grow from near the bone about a quarter of an inch past the base of the nail. This area is referred to as the "nail root." Fingernails grow about an eighth of an inch per month, and toenails grow about a sixteenth of an inch per month. The cuticle forms a protective barrier between the nail and the skin. The nail bed, located under the nail, is richly supplied with blood vessels and is very sensitive. The natural pink color of nails is the result of blood vessels (capillaries) close to the surface of the nail.

The Topical and Internal Uses of Silica In 1993, Dr. A. Lassus, a researcher in Finland, studied women with aged skin, thin hair and brittle nails. These women were given 10 ml of silicic acid once daily for 90 days, and they also applied colloidal silicic acid to their faces for 10 minutes daily. The women who completed the study showed a significant improvement in the quality and appearance of their skin and the condition of their hair and nails.

In 1996, Lassus conducted a double blind study of the

effect of silica on acne and skin oil production. Altogether, 30 patients who had chronic acne of the face (19 females and 11 males whose mean age was 19 years) were divided into two groups. Half were treated topically with a silica product for 20 minutes twice daily for six weeks. The results were statistically significant. After a three-month period, the patients using the silica showed either complete cure or improvement over those who did not use the silica.

Both these studies support the use of oral and topical silica for the improvement of skin, hair and nails. So if you want to put the sheen back into your hair and have strong, healthy nails and clear skin, you need to ensure that your body receives a plentiful supply of nutrients, especially silica.

AIDING CARDIOVASCULAR HEALTH

One of the most common problems of aging in the United States is cardiovascular disease. In fact more than half the adults who die in the United States each year die of heart and blood vessel disease (cardiovascular disease). Atherosclerosis, a condition of coronary arteries, is characterized by an abnormal accumulation of lipid substances and fibrous tissue in the vessel wall. This leads to changes in arterial structure and function and a reduction in blood flow to the heart. It must be noted that unusually high amounts of bound silica are present in the arterial wall.

The relationship between trace elements and human health has been scarcely studied with respect to cardiovascular diseases and hypertension. However, recent findings of a significant number of animal studies demonstrate the role silica plays in delaying the onset and reducing the extension of the atherosclerotic processes.

High blood pressure (hypertension) is an important coronary artery disease risk factor and appears to accelerate the atherosclerotic process. Evidence has been provided by Bataillared et al. (1995) suggesting an association between hypertension and immune dysfunction in Lyon hypertensive (LH) rats. The blood pressure of silica-treated LH rats was lower than that of untreated LH rats one week after treatment began. The effect persisted one week after cessation of the silica treatment. The researchers concluded that weekly administration of silica in young LH rats attenuates the development of hypertension.

The development of atherosclerosis has been related to the silica content of blood. Since silica has been shown to be a component of nucleic acids (DNA, RNA), there is the possibility that silica may influence mutation (genetic changes) in these substances. If that possibility exists, then a connection between silica levels and atherosclerosis would be established. As early as 1958, Loeper and Loeper found that in atherosclerosis artery walls showed an increase in calcium and a significant decrease in silica.

De Francisco and Oliviero (1970) examined the effects of silica in the drinking water of rats with both normal and elevated blood cholesterol levels. The high-cholesterol rats showed lower cholesterol levels when given high-silica drinking water. Experiments with rabbits showed that silica supplementation protected the animals against atherosclerotic plaques (J. Loeper, et al., 1979). Despite the fact that they were fed an unhealthy diet, the rabbits given silica supplements were clearly protected against plaque formation in their arteries.

The role of fiber in the prevention and cure of atherosclerosis in humans has also been documented. Diets high in fiber are typically low in fat and cholesterol and are commonly prescribed for people with all types of cardiovascular disease. (It must be noted that people on high-fiber diets have also been shown to excrete more bile acids, sterols and fat than those on low-fiber diets.) But another factor must be considered in that equation: most fibers are richly supplied with silica. Schwarz (1977) and Loeper et al. (1979) have independently shown that different fibers have varying effects on blood cholesterol. Moreover, both studies confirmed that some fibers are more effective in preventing experimental models of atherosclerosis, reducing cholesterol and blood-lipid levels and binding bile acids in vitro. Exceptionally large amounts of silica (1000 to 25,000 ppm) were found in fiber products of greatly varying origin and chemical composition, and inactive materials, such as different types of purified cellulose, contained only negligible quantities of the element. The findings showed that soluble fibers such as pectin and guar gum have greater cholesterol-lowering effects than the insoluble fibers cellulose and lignin. Indeed, rolled oats and oat bran (rich in soluble fiber and silica) have favorable effects on blood cholesterol, while wheat bran (high in cellulose) is far less effective. In fact two out of three samples of bran had relatively low levels of silica. That could help explain why Schwarz found that bran does not lower serum cholesterol. [The chemical nature of silica in different types of fiber is not well known. It could exist in orthosilicic acid, polymeric silicic acid, colloidal silica (opal), dense silica concentrations or organically bound derivatives of silicic acid (silanolates).]

A logical argument can therefore be made for the hypothesis that the lack of silica may be an important etiological (disease-causing) factor in atherosclerosis. Conversely, silica may well be the active agent in dietary fiber which affects the development of atherosclerosis. Researchers Parr (1980) and Bassler (1978) separately noted that the relatively high silica content of dietary fiber, hard water and wine may contribute to the apparent protection from coronary disease afforded by all these agents.

The fact that there is a low incidence of atherosclerosis in people who live in less developed countries may be related to the availability of dietary silica in those regions. It's widely known that dietary silica is significantly reduced by Western manufacturing processes used to make such foods as white flour and refined soy products. These foods are significantly lower in silica than their respective crude natural products.

A recent finding points to an even more exciting role for silica in promoting cardiovascular health, as well as preventing other degenerative diseases. McCarty (1997) hypothesizes that the production of heparan sulfate proteoglycans (HSPGs), a sulfated mucopolysaccharide, is stimulated by oral administration of glucosamine (an amino derivative of glucose) and thereby exerts a cardiovascular-protective action. McCarty suggests bioavailable silica may likewise increase HSPGs, thus decreasing cardiovascular risk. McCarty speculates that if bioavailable silica supplementation can exert this effect, silica might well complement the activity of glucosamine in the prevention and treatment of osteoarthritis, osteoporosis, atherosclerosis and vascular aneurysms or varicies.

FIGHTING CANCER

Today, more than 2 million cancer patients are being treated in America. It is estimated that cancer will become the leading cause of death in America by the year 2000. The protective action of silica in cases of cancer is suggested by the claim that cancer seldom begins in silica-rich geographic regions. For instance, an early correlation of the protective role of silica was made in Daun, a small town in Germany. In 1932, the spring water of the town was demonstrated to have a rich content of silica and the occurrence and mortality statistic of cancer was low. However, so many other factors are involved in the incidence of cancer that it is doubtful if it could be broadly related to the amount of silica in food or drinking water in a particular region.

Voronkov, Zelchan and Lukevits (1975) postulated that cancer is more frequent in higher organisms because less silica is present as a result of evolutionary processes. Since silica is present in DNA and RNA, the reported changes observed in silica levels of cancer patients warrant and justify further investigation of silica as a possible factor in cancer prevention.

INTERACTION OF SILICA AND ALUMINUM

Aluminum is one of the most abundant minerals on earth, second only to silica, and is found in the earth in its metallic form. However, it is highly toxic to humans in any form. There is no question that the metallic form of aluminum can be leached from aluminum cooking utensils and absorbed from antiperspirants and then deposited in both brain and heart tissue. A highly dangerous, alarmingly pervasive source of aluminum mobilized by acidity in soils, rivers and lakes is responsible for the toxic effect of acid rain on plants and aquatic life. Therefore aluminum is in virtually everything we touch, most of the air we breathe, the water we drink and most foods we eat. The World Health Organization estimates the average adult dietary aluminum intake is between 10 and 15 mg per day.

With the major part of the intake of aluminum being food and antacid medication, an important question is what is the effect of silicic acid in water on the absorption of the aluminum in food. Birchall (1994) conducted a study that showed the antagonistic interactions between silica and aluminum occurring in living organisms. He observed that fish in acidic waters containing aluminum have high mortality due to gill damage and loss of regulatory function. However, in the presence of a high level of silica, aluminum was prevented from binding at gill epithelial surfaces which prevented systemic absorption. This exclusion occurred at the interface between creature and the external environment, but it raises a fundamental question: Is this a general effect that takes place not only at the fish gill but also at plant root membranes and in the gastrointestinal tract of mammals and humans?

In another study Forbes and Agwani (1994) showed that silica abolished the acute toxicity of aluminum in fish. They postulated therefore that silica would significantly reduce the gastrointestinal absorption of aluminum in humans by forming compounds which could not be absorbed.

Other early experiments indicate that the inhibitory effects of aluminum are absent in the presence of silicic acid. Seaborn and Nielsen (1994) performed an experiment to ascertain whether high dietary aluminum would accentuate the signs of silica deprivation in rats and conversely whether silica deprivation would accentuate their response to high dietary aluminum. The findings indicate that high dietary aluminum in rats can affect their response to silica deprivation and dietary silica can affect their response to high dietary aluminum.

The biological significance of the reaction of silica and aluminum is of particular and urgent interest in the field of human neurodegenerative diseases, especially Alzheimer dementia. (Alzheimer's disease usually begins in later middle life as a slight defect in memory. The person is confused, restless and unable to carry out purposeful activities. The disease worsens progressively as a person ages.) Bellia et al. (1996) gave volunteers orange juice containing aluminum with and without silica. A significant reduction in aluminum absorption in the gastrointestinal tract was caused by silica, even in the presence of citrate, which is known to enhance the absorption of aluminum.

PREVENTING ALZHEIMER DEMENTIA

The possible involvement of environmental aluminum in the etiology of Alzheimer disease is supported by findings of the abnormal presence of aluminum in human and animal brain tissues. A preliminary study reported by Birchall (1994) was undertaken to investigate the effect of dietary silica and aluminum on levels of the two minerals found in the brain. Regional variations in silica, which were independent of dietary silica supplementation, suggest that silica may be an essential element in the brain. Aluminum supplementation decreased the silica content in selected brain regions, including those thought to be involved in Alzheimer's disease.

Birchall established a relationship among silica, aluminum and age. In 23-month-old rats, aluminum supplementation did not increase brain aluminum content. By contrast, in 28-month-old rats, aluminum supplementation with a low silica diet increased brain aluminum content in most regions. No increase occurred in silica-supplemented groups of the same age. Dietary silica supplementation thus appeared to be protective against aluminum accumulation in the aging brain.

According to Birchall (1994), researchers have investigated the geographic correlation between Alzheimer's disease and levels of aluminum in water supplies. They found that aluminum contents are high when silica is low. Using isotopic aluminum administered orally to five healthy volunteers in the presence and absence of silica, researchers found silica reduced the peak plasma aluminum concentration to 15 percent of the value obtained in the absence of silica. The results indicate that dissolved silica is an important factor in limiting the absorption of dietary aluminum.

Forbes and Agwani (1994) have shown that relatively high risks of mental impairment are frequently associated with relatively low levels of silica in drinking water. Such effects may be less likely at higher silica concentrations. Generally, the various associations with a measure of mental impairment can be explained by the assumption that silica exerts a protective role against the biotoxic effects of aluminum. Jacqmin-Gadda et al. (1996) investigated the silica and aluminum content of drinking water and cognitive impairment in the elderly. The researchers found that the association between cognitive impairment and aluminum depended on the pH and concentration of silica in drinking water. High levels of aluminum appeared to have a deleterious effect when silica concentration was low, but there was a protective effect when the pH and silica levels (10.4 mg silica liter) were high. The researchers found a complex association between components of drinking water and cognitive impairment. Among nine elements of drinking water studied, only pH and the concentration of aluminum, calcium and silica were found to be associated with cognitive impairment. They concluded that a high aluminum content of drinking water may be associated with a high risk of cognitive impairment only when the concentration of silica is low.

In 1994, Fasman and Moore working at Brandeis University showed the effect of sodium orthosilicate in reversing the damaging effects of aluminum on brain proteins of Alzheimer's patients. These results provide additional information toward understanding the role of aluminum in the Alzheimer-diseased brain and suggest the investigation of the possible use of silicates as a therapeutic agent. Anything that could reduce the rapidly accelerating rise of this vicious degenerative disease would be a much-heralded breakthrough. Wouldn't it be astounding if prevention were as simple and readily available as silica supplementation?

STOPPING THE AGING PROCESS

Many people today turn to nutritional supplements of DHEA, melatonin and specific herbs to help slow the body's aging process. The widespread popularity of antioxidant nutrition is a primary example of people's deep-rooted interest in maintaining youth and vitality for as long as possible. We know, based on the oldest recorded living woman who died in 1997 at age 122, that we can expect to live a maximum of 122 years. Most people would agree that they would like to live a maximum number of good quality years. Although reports today indicate an average life expectancy of 76 years, people are constantly trying to extend that projection. A recent study of women having children in their 40s demonstrated a positive correlation with longer life.

Numerous theories of aging have been advanced from the time of Ponce de Leon's claim that he found the fountain of youth in the 16th century. Some believe that the body simply wears out after years of wear and tear like the parts of a car. Others consider the loss of body function in aging as a loss of cellular hydration. How and why the body ages has been the subject of many books and postulations. No conclusive data exist, but one thing is certain: Nutrition definitely plays a role in aging.

Fregert (1958) reported that the silica content of the human heart declined with age. By 40 the human heart has half the silica content it had in infancy. It has also been reported that the silica content of heart tissue decreases with increased sclerotic damage. That type of finding has led to the idea of a therapeutic or preventive role for silica.

Silica has been considered to inhibit the aging process in tissues. Consider the flexibility of a baby or young child whose bodies are richly saturated with silica. As we grow older we have less silica in our bodies and our flexibility is also lost. This is due in part in Western industrialized countries, where people eat a diet high in processed foods, to a lack of silica replacement from the food supply. Theoretically, silica may play a definite role in the prevention of premature aging. Considering the role silica plays in maintaining the youthful appearance of hair, skin and nails and its many valuable functions in disease prevention, it appears that silica should be looked at more seriously as an essential trace element needed for maintaining the youth and vitality of the whole body.

SAFETY: INTAKE CONSIDERATIONS

Considering the universality of silica in nature, it is not surprising that it is considered harmless in food and drink. Evidence of its nontoxicity is the observation that magnesium trisilicate, an over-the-counter antacid, has been used by humans for more than 40 years without obvious deleterious effects. Other silicates are used as food additives. In 1980, toxicity studies of silica gel were conducted in Germany using laboratory mice. Animals were fed a varying range of silica gel over a period of two weeks. The gel was demonstrated to be nontoxic even at high concentrations.

Taken orally, silica is at worst almost certainly innocuous. Rabbits have been fed 2 grams of silica gel per day for 30 weeks with no ill effects (Gye & Purdy, 1924). Magnesium silicate, which is easily attacked by acid, thus liberating the silica, is widely used in internal medicine.

However, when silica enters the body by inhalation, a type of toxicity almost always results. Most of the negative literature about silica cites inadvertent inhalation of silica dust. The silica retained by the lungs leads to the condition of silicosis. *It must be emphasized that the context in which silica is discussed in this text refers exclusively to the oral ingestion of silica as a dietary supplement.*

Silica has no known level of toxicity when ingested orally. The average individual already consumes 20 to 60 mg silica daily from food and drink. Significant differences certainly exist in the metabolism of each individual and, in particular, a person's ability to retain silica in the body. However, there is also evidence that silica is rapidly lost from the body, which raises some concerns about the adequacy of silica intake by the average individual. Carlisle's research also demonstrates that the level of measurable silica in the body decreases with age. As substantiated by the evidence discussed throughout this text, silica supplementation has been shown to help numerous degenerative problems that are commonly seen in the population. In light of silica's safety record and silica's role in degenerative disease prevention, it would be fair to say that a dosage of 300 to 500 mg daily represents the low end of the maintenance dosage range. Considering a person's individual metabolic need for silica, levels of intake may even be higher.

Individuals plagued with numerous problems of a degenerative nature may look at a silica dose of 700 to 1,000 mg daily. Rabbits have been given very high silica doses over a period of 30 weeks with virtually no ill effects. This is certainly a situation where more may very possibly be better than less. While researchers have demonstrated the requirement for silica in our diet, it is apparent that minimum and maximum levels of intake also need to be established. Research studies are warranted to determine these levels in order to establish effective levels of silica intake based on scientific evidence.

CONCLUDING THOUGHTS

The therapeutic effects of hot mineral springs have long been known throughout Europe. Native American Indians have a tradition of using silica-rich mud for healing wounds and as a basis for poultices. While the specific qualities of individual minerals was not truly understood for many centuries, and many are still not thoroughly understood, the frequent users of mineral spas fully understood their beneficial effects. They help alleviate symptoms of arthritis, tuberculosis and muscle aches and pains, to name just a few of their many benefits.

Though it's not widely known, President Franklin D. Roosevelt rebuilt a famous spa in Georgia for the express purpose of rehabilitation from polio. While he was not overly successful in curing the problem, he certainly received great physical and mental relief from using the spa, as did everyone else fortunate enough to spend time there.

At the end of World War II, Dr. Becker in Germany began work on creating a silica-based therapeutic. Before 1950, he had developed a silica gel which proved very useful as an anti-inflammatory. In the late 1940s and early 1950s a method of extracting silica from horsetail was developed and patented. This new method used only water and produced a nontoxic bioavailable silica. This development took science and the nutritional supplement industry to the next level of research on the nutritional effects of silica.

Silica has still not received the proper recognition required for widespread use as an essential mineral supplement. In fact, it would be surprising if any more than 1 percent of the population at large had any concept of the benefits of silica. However, the many benefits of silica supplementation, whether it be mineral or vegetal, are slowly but surely gaining recognition. Although much of the research on silica has focused on connective tissue and bone, it is clear that silica serves many other vital functions in the body.

From the vision of Pasteur to the research work of Carlisle, Iler and Schwarz, researchers today are continuing to demonstrate positive findings for silica. It is clear from the information presented by the scientific community that the most bioavailable form of silica is silicic acid. Colloidal silica supplements found in the natural products market can be ingested safely, and silica lotion products can be applied to the skin directly, without concern for both toxicity and adverse reactions. When applied topically, silica is effectively absorbed and utilized. One of the advantages of colloidal silica supplements is the ease with which they can be taken. These liquid products provide silicic acid in purified water. Since particle size of silicic acid is so small, colloidal silica provides a maximized delivery system that allows the mineral to carry out its many essential functions in the body.

For some as yet unknown reason, aging is associated with a decrease in the silica content of the body. This observation has been interpreted as an indication of why we need to consider silica supplements as the years mount up. This has led many to believe silica has a preventive role in premature aging. Consider the research by Lassus (1996), which demonstrates that dry, wrinkled skin which has lost its hydrated properties and elasticity dramatically improves in appearance when silica is applied topically.

Beyond that observation, one needs to understand the other significance of this change in the silica content in the body. It has been demonstrated that mineral elements interact, sometimes causing a binding of minerals which renders them unavailable. Also, dietary data clearly show the amount of silica we ingest without supplementation to be small and perhaps inadequate to meet the physiologic challenges of the body as it ages. Considering the research data, today's average diet only provides enough silica to meet basic needs in connective tissue formation, which is minimally met by dietary silica intake alone.

Another interesting observation is the presence of silica in fiber sources, which clearly demonstrates a cardiovascular protective role. While it is highly recommended that people continue to improve the fiber content of their diet, the silica component of fiber has thus far been overlooked as the active agent in fiber's cardiovascular protective role. The research shows that while silica has a direct, positive effect in cardiovascular health, perhaps it also has an indirect or overlooked effect as a fiber component. How many other overlooked roles and effects are there for silica? Only continued research on this fascinating mineral will reveal them.

Silica research is still in its infancy. The final chapter on silica has clearly not been written. Many novel functions for silica will surely be discovered as its many miraculous powers are explored. The results of silica supplementation are already remarkable. Testimonials from people using colloidal silica liquid supplement daily demonstrate an amazing array of positive results. External body improvements of the hair, skin and nails show the many positive cosmetic roles of silica, but its value goes much deeper. Research studies and testimonials of colloidal silica supplement users clearly demonstrate the valuable role of this essential nutrient in improving many degenerative diseases and restoring overall health. Silica is truly nature's crystal building block.

REFERENCES

Baker, et al. 1961. See Iler.

Bassler, T.J. 1978. "Hard water, food fibre and silicon." *Br. Med. J.* 1: 919.

Bataillard, A., Renaudin, C., Sassard, J. 1995. "Silica attenuates hypertension in Lyon hypertensive rats." *J. Hypertens.* 13: 1581-84.

Bellia, J.P., Birchall, J.D., Roberts, N.B. 1996. "The role of silicic acid in the renal excretion of aluminum." *Ann. Clin. Lab. Sci.* 26(3): 227-33.

Bergna, H. 1994. *The Colloidal Chemistry of Silica.* Advances in Chemistry Series 234. Washington, D.C.: American Chemical Society.

Bertone, A.L. 1989. "Principles of wound healing." *Vet. Clin. NA Equine Prac.* 5: 449-63.

Birchall, J.D. 1994. "Silicon-aluminum interactions and biology" in Bergna, H., ed. *The Colloidal Chemistry of Silica.* Washington, D.C.: American Chemical Society, pp. 601-15.

—————. 1995. "The essentiality of silicon in biology." *Chem. Soc. Rev.* 24: 351-57.

—————, Espie, A.W. 1986. "Biological implications on the interaction (via silanol groups) of silicon with metal ions." *Ciba Found. Symp.* 121: 140-59.

Bolyshev. 1952. See Iler.

Burton, A.C. 1980. "Protection from calcium by silica in the water supply of U.S. cities." *J. Environ. Path. Tox.* 4: 31-40.

Carlisle, E.M. 1970. See Iler.

————. 1972. "Silicon: an essential element for the chick." *Science* 178: 619-21.

————. 1974. "Silicon as an essential element for the chick." *Fed. Proc.* 33: 1758-66.

————. 1981. "Silicon in bone formation" in Simpson, T.L., Volcani, B.E., eds. *Silicon and Siliceous Structures in Biological Systems*. New York: Springer, pp. 69-94.

————. 1984. "Silicon" in Frieden, E., ed. *Biochemistry of the Essential Ultratrace Elements*. New York: Plenum Press, pp. 257-91.

————. 1988. "Silicon as a trace nutrient." *Sci. Tot. Environ.* 73: 95-106.

Charnot, Y., Peres, G. 1971. "Contribution a l etude de la regulation endocrinienne du metabolisme silicique." *Ann. Endocrin.* 32: 397-402.

DeFrancisco and Oliviero. 1970. See Iler.

Draper and Bell. 1980. See Seaborn and Neilsen.

Edwardson, J.A., et al. 1993. "Effect of silicon on gastrointestinal absorption of aluminum." *Lancet* 342: 211-12.

Eitel, F., Sklarek, J. 1988. "Wound healing and wound dressing." *Tierarztl Prax.* 16: 1-12.

Eliot, M.A., Edwards, H.M. 1991. "Effect of dietary silicon on growth and skeletal development in chickens." *J. Nutr.* 121: 201-07.

Fasman and Moore. 1994. See Bergna.

Forbes, W.F., Agwani, N. 1994. "A suggested mechanism for aluminum biotoxicity." *J. Theor. Biol.* 171: 207-14.

————, McLachlan, D.R. 1996. "Further thoughts on the Alzheimer's disease link." *J. Epidemiol. Comm. Health* 50(4): 401-03.

Fregert. 1958. See Iler.

Frison. 1948. See Iler.

Gerstein, A.D., et al. 1993. "Wound healing and aging." *Derm. Clin.* 11: 749-57.

Gye, Purdy. 1924. See Iler.

Hidaka, S., Okamoto, Y., Abe, K. 1993. "Possible regulatory roles of silicic acid, silica and clay minerals in the formation of calcium phosphate precipitates." *Arch. Oral Biol.* 38: 405-13.

Houtman, J.P. 1996. "Trace elements and cardiovascular diseases." *J. Cardio. Risk.* 3: 18-25.

Iler, R.K. 1979. *The Chemistry of Silica: Solubility, Polymerization, Colloid and Surface Properties, and Biochemistry*. New York: Wiley-Interscience Publication.

Jacqmin-Gadda, H., et al. 1996. "Silica and aluminum in drinking water and cognitive impairment in the elderly." *Epidemiology* 7: 281-85.

Jones, Handreck. 1965. See Iler.

Kaufmann, K. 1995. *Silica: The Amazing Gel*. Canada: Alive Books.

LaHeu. 1970. See Iler.

Lassus, A. 1996. "The effect of silicol gel compared with placebo on papulopustular acne and sebum production. A double blind study." *J. Int. Med. Res.* 24(4): 340-44.

Lewin. 1957, 1958. See Iler.

Loeper, J., Loeper, J., Fragny, M. 1979. "The physiological role of the silicon and its antiatheromatous action" in Bendz, G., Linquist, I., eds. *Biochemistry of Silicon and Related Problems*. New York: Plenum, pp. 281-96.

Lundie. 1913. See Iler.

Mancinella, A. 1991. "Silicon a trace element essential for living organisms. Recent knowledge on its preventive role in atherosclerotic process, aging and neoplasms." *Clin. Ter.* 137: 343-50.

McCarty, M. 1997. "Glucosamine may retard atherogenesis by promoting endothelial production of HSPGs." *Med. Hypoth.* 48: 245-56.

—————. 1997. "Reported antiatherosclerotic activity of silicon may reflect increased endothelial synthesis of heparan sulfate proteoglycans." *Med. Hypoth.* 49: 175-76.

Najda, J., et al. 1992. "Silicon metabolism. The interrelations of inorganic silicon with systemic iron, zinc, and copper pools in the rat." *Biol. Trace Elem. Res.* 34: 185-95.

Nielsen, F.H. 1991. "Nutritional requirements for boron, silicon, vanadium, nickel and arsenic: current knowledge and speculation." *FASEB J.* 5: 2661-67.

Nouaigui, H., et al. 1989. "Invasive and noninvasive studies of the protective action of a silicon containing cream and its excipient in skin irritation induced by sodium laurylsulfate." *Ann. Derm. Venereol.* 116: 389-98.

Oberbaum, M., et al. 1992. "Wound healing by homeopathic silica dilutions in mice." *Harefuah.* 123: 79-82.

Parr, R. 1980. "Silicon, wine and the heart." *Lancet* i: 1087.

Pennington, J.A. 1991. "Silicon in foods and diets." *Food Addit. Contam.* 8: 97-118.

Reimann. 1965. See Iler.

Schroeder. 1973. See Iler.

Schwarz, K. 1974. "Recent dietary trace element research exemplified by tin, fluorine and silicon." *Fed. Proc.* 33: 1748-57.

—————. 1977. "Silicon, fibre and atherosclerosis." *Lancet* i: 454-57.

Schwarz and Milne. 1972. See Iler.

Seaborn, C.D., Nielsen, F.H. 1993. "Silicon: A nutritional beneficience for bones, brain and blood vessels?" *Nutr. Today* July/Aug.: 13-18.

—————, —————. 1994. "High dietary aluminum affects the responses of rats to silicon deprivation." *Biol. Trace Elem. Res.* 41: 295-304.

Siever. 1957. See Iler.

Sprecher. 1911. See Iler.

Sreenivasan. 1934. See Iler.

Stone and Gray. 1948. See Iler.

Uber, C.L., McReynolds, R.A. 1982. "Immunotoxicology of silica." *Curr. Rev. Tox.* 10: 303-19.

Volkenheimer. 1964. See Iler.

Voronkov, Zelchan, Lukevits. 1975. See Iler.

Wagner. 1940. See Iler.

Werner. 1968. See Iler.

Zatta, P., et al. 1988. "Alzheimer's dementia and the aluminum hypothesis." *Med. Hypotheses* 26: 139-42.

Zitelli, J. 1987. "Wound healing for the clinician." *Adv. Derm.* 2: 243-67.